I0797661

CLOWN FISH

by Emma Bassier

Cody Koala
An Imprint of Pop!
popbooksonline.com

abdobooks.com
Published by Pop!, a division of ABDO, PO Box 398166, Minneapolis, Minnesota 55439.

Printed in the United States of America, North Mankato, Minnesota

052019
092019

THIS BOOK CONTAINS RECYCLED MATERIALS

Cover Photo: iStockphoto
Interior Photos: iStockphoto, 1, 5 (top), 5 (bottom left), 5 (bottom right), 7 (top), 7 (bottom left), 7 (bottom right), 9, 10, 13, 16, 19; Louise Murray/Science Source, 14; Scubazoo/Science Source, 20

Editor: Meg Gaertner
Series Designer: Jake Nordby

Library of Congress Control Number: 2018964496

Publisher's Cataloging-in-Publication Data

Names: Bassier, Emma, author.
Title: Clown fish / by Emma Bassier.
Description: Minneapolis, Minnesota : Pop!, 2020 | Series: Ocean animals | Includes online resources and index.
Identifiers: ISBN 9781532163371 (lib. bdg.) | ISBN 9781644940105 (pbk.) | ISBN 9781532164811 (ebook)
Subjects: LCSH: Clown fishes--Juvenile literature. | Anemonefishes--Juvenile literature. | Marine fishes--Juvenile literature. | Ocean animals--Juvenile literature.
Classification: DDC 597.72--dc23

Hello! My name is

Cody Koala

Pop open this book and you'll find QR codes like this one, loaded with information, so you can learn even more!

Scan this code* and others like it while you read, or visit the website below to make this book pop.

popbooksonline.com/clown-fish

*Scanning QR codes requires a web-enabled smart device with a QR code reader app and a camera.

Table of Contents

Chapter 1

Bright Orange Body

Clown fish are small ocean fish. They are bright orange and have three white stripes. Their **fins** help them swim.

Watch a video here!

Chapter 2

Deadly Homes

Clown fish live in colorful **sea anemones**. Sea anemones use their **tentacles** to sting and hurt fish. But their stings do not hurt clown fish.

Learn more here!

Clown fish are covered in **mucus**. Scientists believe the layer of mucus protects them. It keeps them safe from the sea anemones' stings.

Clown fish help sea anemones by keeping them clean.

white stripe
fin
tentacle

Clown fish find a sea anemone to live in. They brush their bodies against the sea anemone's tentacles. The clown fish and sea anemone get used to one another. Then the clown fish can enter their new home.

Chapter 3

Close to Home

Clown fish live in warm water. They live in the Indian Ocean. They are also found in the western Pacific Ocean.

Learn more here!

Most clown fish stay close to their **sea anemone** for their whole lives. They eat small plants and animals off the sea anemone's **tentacles**.

Clown fish sometimes leave the sea anemone to find more food. They swim around. They look for tiny creatures called plankton.

Chapter 4

In the School

All clown fish are born **male**. Clown fish swim in groups called schools. Each school of clown fish lives together in the same **sea anemone**.

Complete an activity here!

eggs

The largest male clown fish in the school becomes a **female**. The female lays her eggs. She and the next-largest male protect the eggs. Then the eggs hatch.

Clown fish usually live for six to ten years.

Making Connections

Text-to-Self

Have you ever seen a clown fish in real life? If yes, what did you think of it? If no, would you want to?

Text-to-Text

Have you read other books about ocean animals? How are those animals similar to or different from clown fish?

Text-to-World

Clown fish live in sea anemones. Can you think of two other animals that live together?

Glossary

female – a person or animal of the sex that can have babies or lay eggs.

fin – a thin, flat limb that helps fish and other water animals swim.

male – a person or animal of the sex that cannot have babies or lay eggs.

mucus – a slimy substance released by an animal's body.

sea anemone – an ocean animal that uses its tentacles to sting and grab fish for food.

tentacle – a flexible, moving limb.

Index

Online Resources

popbooksonline.com

Thanks for reading this Cody Koala book!

Scan this code* and others like it in this book, or visit the website below to make this book pop!

popbooksonline.com/clown-fish

*Scanning QR codes requires a web-enabled smart device with a QR code reader app and a camera.